# QCE Biology

Revision Workbook Answers

*greenhill

Copyright © Gemma Dale 2024

All rights reserved. No part of this publication may be reproduced, distributed, or transmitted in any form or by any means, including photocopying, recording, or other electronic or mechanical methods, without the prior written permission of the publisher, except in the case of brief quotations embodied in critical reviews and certain other non-commercial uses permitted by copyright law.

**✱ green**hill

https://greenhillpublishing.com.au/

Dale, Gemma (author)

*QCE Biology: Revision Workbook Answers*

ISBN 978-1-923333-47-5

TEXTBOOK | BIOLOGY

Typesetting Forma DJR 10.5/16

Cover and book design by Green Hill Publishing

Diagrams by Matilda Crossley

# QCE
# Biology
## Revision Workbook Answers

**DR GEMMA DALE**

ILLUSTRATED BY
**MATILDA CROSSLEY**

# CONTENTS

## Unit 1

**Cells and multicellular organisms** — 1

**Topic 1: Cells as the basis of life** — 2
Section 1: Cell Structure — 2
Section 2: Cell Division — 4
Section 3: Cell Membrane — 6
Section 4: Cell Transport — 7

**Topic 2: Exchange of nutrients and wastes** — 9
Section 1: Biochemistry — 9
Section 2: Enzymes — 12
Section 3: Exchange Surfaces — 15
Section 4: Excretory System — 17

**Topic 3: Cellular energy, gas exchange and plant physiology** — 19
Section 1: Respiration — 19
Section 2: Photosynthesis — 21
Section 3: Gas Exchange — 22
Section 4: Plant Physiology — 23

## Unit 2

**Maintaining the internal environment** — 27

**Topic 1: Homeostasis - thermoregulation and osmoregulation** — 28
Section 1: Nervous System — 28
Section 2: Endocrine System — 31
Section 3: Thermoregulation — 32
Section 4: Osmoregulation — 34

**Topic 2: Infectious disease and epidemiology** — 36
Section 1: Disease — 36
Section 2: Immune Response — 38
Section 3: Epidemiology — 42

# Unit 1

# Cells and multicellular organisms

# TOPIC 1: CELLS AS THE BASIS OF LIFE

## Section 1: Cell Structure

### Question 1
Ribosomes; DNA. (2)

### Question 2
Mitochondria: (statements should be comparative) have a double membrane; no cell wall; no capsule, no flagellum/pili; no plasmids. (2)

### Question 3
A; mitochondrion – produces ATP in aerobic respiration
B: rough endoplasmic reticulum – synthesises proteins
C: nucleus – stores DNA, controls cell activities (3)

### Question 4

| Prokaryotic cells | Eukaryotic cells |
| --- | --- |
| No nucleus | Have a nucleus |
| No membrane-bound organelles | Have membrane-bound organelles |
| 70S ribosomes | 80S ribosomes |
| Smaller (0.1-5 µm) | Larger (10-100 µm) |

### Question 5

| Feature | Prokaryotic cell | Eukaryotic cell |
| --- | --- | --- |
| Cell membrane | √ | √ |
| Nucleus |  | √ |
| Ribosomes | √ | √ |
| Rough endoplasmic reticulum |  | √ |
| Mitochondria |  | √ |

### Question 6
Lysosomes; flagella; cilia

### Question 7
Storage of water, nutrients and waste products; Maintaining cell structure

# Section 1: Cell Structure

### Question 8
They contain enzymes that: break down old or damaged organelles; digest pathogens that enter the cell.

### Question 9
Nucleus, chloroplast, mitochondrion

### Question 10
a) W: rough endoplasmic reticulum

X: Golgi body/apparatus

Y: Nucleus

Z: Mitochondrion

b) They require energy to secrete the mucus.

c) It modifies mucus proteins, places them in vesicles and then sends these to the plasma membrane to be released

### Question 11

| Organelle | Name | Function |
|---|---|---|
| A | Chloroplast | Photosynthesis (in plants) |
| B | Ribosomes | Synthesises proteins |
| C | Nucleus | Controls cell activities, containing DNA |
| D | Rough endoplasmic reticulum | Synthesises and processes proteins |
| E | Cytoplasm | Reactions within cells often take place within the cytoplasm. |
| F | Vacuole | Stores nutrients and waste; maintains cell structure (in plants) |
| G | Cell wall | Provides structure and support (in plants) |
| H | Cell (plasma) membrane | Controls the movement of substances in and out of the cell |
| I | Nucleolus | Ribosome synthesis |
| J | Mitochondrion | Site of aerobic respiration, producing ATP |
| K | Golgi body | Modifies, packages and transports proteins and lipids |
| L | Smooth endoplasmic reticulum | Synthesises lipids and steroids; metabolises carbohydrates; regulates calcium; detoxifies drugs |

# Unit 1: Topic 1: Cells as the basis of life

## Section 2: Cell Division

### Question 1
An unspecialised cell that can divide indefinitely, self-renew through continuous cell division, and ultimately differentiate into all specialised cells.

### Question 2
Can give rise to all cell types in an organism, including those needed to form an entire organism, including the generation of all embryonic and extra-embryonic tissues, such as those in the placenta.

### Question 3
Stem cells from the heart are usually multipotent, meaning they can only develop into heart tissue and no other types of cell. They are specialised for specific developmental pathways that produce heart tissue and therefore unable to repair skin.

### Question 4
Interphase

### Question 5
A tissue is a group of similar cells working together to perform a specific function, like muscle or epithelial tissue. In contrast, a system is composed of various organs and tissues that collaborate to carry out complex functions for the body, such as the circulatory or digestive system.

### Question 6
a) Anaphase
b) Telophase
c) Metaphase
d) Prophase

### Question 7
It is incorrect because the cell is highly active during this phase, engaged in processes such as growth, DNA replication and protein synthesis.

### Question 8
Growth, cell replacement, asexual reproduction.

### Question 9

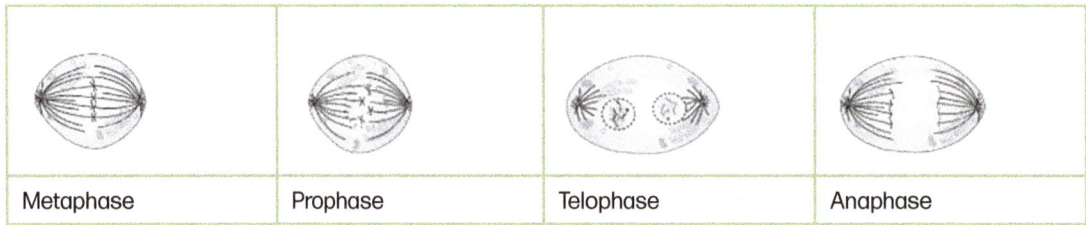

| Metaphase | Prophase | Telophase | Anaphase |

# Section 2: Cell Division

## Question 10
A group of similar cells that work together to perform a specific function.

## Question 11
Prophase; centromere; envelope/membrane; replicated; equator/metaphase plate; anaphase; spindle fibres; genetically.

## Question 12
Mitochondria, as they produce the ATP required for the liver cells' metabolic functions.

## Question 13
A structure composed of different types of tissues working together to perform a specific function.

## Question 14

|  | Number of chromosomes | Mass of DNA/arbitrary units |
| --- | --- | --- |
| At prophase | 16 | 10 |
| At telophase | 16 | 5 |
| From a sperm cell | 8 | 2.5 |

## Question 15
Embryonic stem cells can treat a broader range of disorders because they are pluripotent, meaning they can develop into any cell type in the body. In contrast, menstrual blood stem cells are typically multipotent, which limits their ability to differentiate into only a few types of cells.

## Question 16

| Stage | Process |
| --- | --- |
| G1 | Growth |
| S | DNA replication |
| G2 | Growth, preparation for mitosis |
| M | Mitosis |

## Question 17
Suggested:

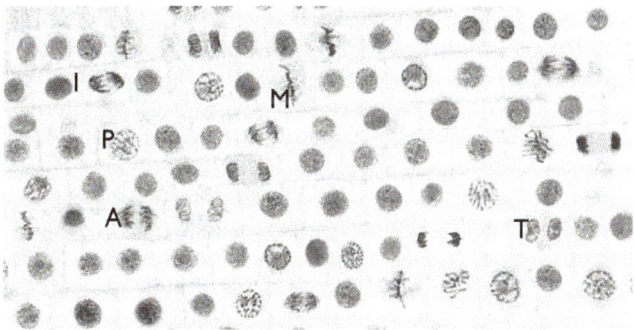

# Section 3: Cell Membrane

### Question 1
a) Fluid mosaic model
b) A: Glycoprotein; B: Carbohydrate chain/oligosaccharide; C: Glycoprotein; D: Channel protein; E: Carrier protein; F: Phospholipid; G: Peripheral protein; H: Cholesterol

### Question 2
Phospholipid molecules form a bilayer because their hydrophilic heads are attracted to water and face outward towards the aqueous environments inside and outside the cell. Meanwhile, their hydrophobic tails repel water and face inward, away from the water.

### Question 3
A phospholipid consists of a hydrophilic (water-attracting) head and two hydrophobic (water-repelling) tails. The head is composed of a phosphate group and a glycerol molecule, while the tails are made of long fatty acid chains.

### Question 4
Cholesterol helps to maintain the fluidity and stability of cell membranes.

### Question 5
a) Ethanol disrupts the phospholipid bilayer of both the cell membrane and the vacuole membrane. The organic solvent dissolves the lipid components of these membranes, compromising their integrity and creating gaps. As a result, the betacyanin pigment, which is normally contained within the vacuole, can escape.
b) Higher temperatures increase the fluidity of the phospholipid bilayers in both the cell membrane and vacuole membrane. This makes the membranes more permeable, allowing the betacyanin pigment to leak out faster.

# Section 4: Cell Transport

## Question 1
a) Difference: Active transport requires energy (usually from ATP) to move substances against their concentration gradient, while facilitated diffusion does not require energy and moves substances down their concentration gradient.
Similarity: Both active transport and facilitated diffusion involve the use of specific transport proteins to help substances cross the plasma membrane.

b) Vitamin C: As a water-soluble vitamin, Vitamin C cannot easily pass through the hydrophobic (water-repelling) interior of the phospholipid bilayer. Instead, it uses specific transport proteins or channels embedded in the plasma membrane to facilitate its entry into the cell.
Vitamin D: Being lipid-soluble, Vitamin D can dissolve in the hydrophobic interior of the phospholipid bilayer. This allows it to diffuse directly through the membrane without the need for transport proteins.

## Question 2
Active transport; endocytosis; exocytosis

## Question 3

|  | Initial movement of water | | Cell bursts? |
|---|---|---|---|
|  | Into cell | Out of cell |  |
| Plant cell in distilled water | ✓ |  |  |
| Animal cell in concentrated salt solution |  | ✓ |  |
| Animal cell in distilled water | ✓ |  | ✓ |

## Question 4
a) X: Rough endoplasmic reticulum; Y: Golgi body/apparatus

b) Z: Exocytosis

c) The rough endoplasmic reticulum (RER) is involved in the synthesis of extracellular enzymes. It has ribosomes on its surface that produce these enzymes, which are then folded and modified within the RER. Once the enzymes are properly processed, they are transported in vesicles from the RER to the Golgi apparatus.

The Golgi apparatus further modifies, packages, and sorts these enzymes. It adds final chemical modifications and then encases them in vesicles for transport to the cell membrane. Finally, the vesicles fuse with the membrane, releasing the enzymes into the extracellular space.

## Question 5
a) Diffusion is a passive process where molecules move from an area of higher concentration to an area of lower concentration until equilibrium is reached.

b) Red blood cells have a biconcave shape, which increases their surface area-to-volume ratio, enhancing the diffusion of oxygen. Additionally, they lack a nucleus, allowing more room for haemoglobin, the protein responsible for binding and transporting oxygen.

# Unit 1: Topic 1: Cells as the basis of life

## Question 6

a) Cube 1: 6:1; Cube 2: 3:1; Cube 3: 2:1

b) Using the results, the surface area-to-volume ratio decreases as the cube size increases: As organisms get larger, their surface area-to-volume ratio decreases, making diffusion less efficient. Multicellular organisms need transport systems because their low surface area-to-volume ratio means diffusion alone cannot efficiently supply all cells with nutrients or remove wastes. In contrast, unicellular organisms have a higher surface area-to-volume ratio, allowing diffusion to be sufficient for their needs.

## Question 7

Exocytosis involves substances leaving the cell, whereas endocytosis involves them entering the cell. Exocytosis involves vesicles fusing with the plasma membrane, whereas endocytosis involves the formation of vesicles from the plasma membrane.

# TOPIC 2: EXCHANGE OF NUTRIENTS AND WASTES

## Section 1: Biochemistry

### Question 1
a) Glycerol and three fatty acids

b) Condensation

c) Energy storage

### Question 2
Amylose is a linear polymer of glucose molecules linked by α-1,4 glycosidic bonds, forming a helical structure. Amylopectin, on the other hand, is a branched polymer where glucose molecules are linked by both α-1,4 and α-1,6 glycosidic bonds. The α-1,6 bonds create branches at intervals along the amylopectin chain, making it more complex and branched compared to amylose.

### Question 3
a) Ester

b) Hydrolysis

### Question 4

| Carbohydrates | Lipids | Proteins |
|---|---|---|
| Proteins | Carbohydrates | Lipids |

# Unit 1: Topic 2: Exchange of nutrients and wastes

## Question 5
a)

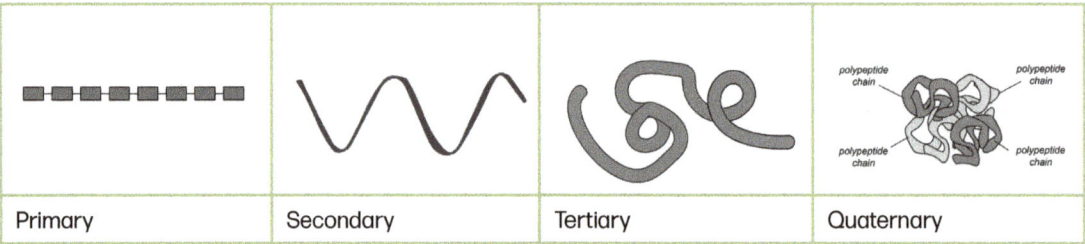

| Primary | Secondary | Tertiary | Quaternary |

b)   Amino acid

## Question 6
A polysaccharide found in animals is glycogen. Its function is to store energy in the liver and muscles, where it can be rapidly broken down into glucose when the body needs a quick energy source. a polysaccharide found in animals and state its function.

## Question 7
They are not composed of long chains of repeating monomer units. Instead, they are made up of a glycerol molecule bonded to three fatty acids.

## Question 8
Triglycerides have a glycerol backbone with three fatty acids attached, whereas phospholipids have the same glycerol backbone but with only two fatty acids and a phosphate group attached.

## Question 9
Saturated fatty acids have no double bonds between the carbon atoms in their hydrocarbon chain, meaning all carbon atoms are fully bonded to hydrogen. This structure allows them to be solid at room temperature. In contrast, unsaturated fatty acids have one or more double bonds between carbon atoms, which creates kinks in the chain. These kinks prevent tight packing, making unsaturated fatty acids typically liquid at room temperature.

## Question 10

| Element | Carbohydrates | Proteins | Lipids |
|---|---|---|---|
| Carbon (C) | ✓ | ✓ | ✓ |
| Hydrogen (H) | ✓ | ✓ | ✓ |
| Oxygen (O) | ✓ | ✓ | ✓ |
| Nitrogen (N) |  | ✓ |  |
| Phosphorus (P) |  |  | ✓ |
| Sulfur (S) |  | ✓ |  |

## Question 11
Animal: Glycogen; Plant: Starch

## Question 12
The tertiary structure of a protein refers to its overall three-dimensional shape. It results from the folding of the protein's polypeptide chain into a compact, globular form. This folding is driven by various interactions between the side chains (R groups) of the amino acids, including hydrogen bonds, ionic bonds, disulfide bonds, and hydrophobic interactions. These interactions stabilise the protein's structure, allowing it to perform its specific biological functions.

# Section 2: Enzymes

## Question 1
a)

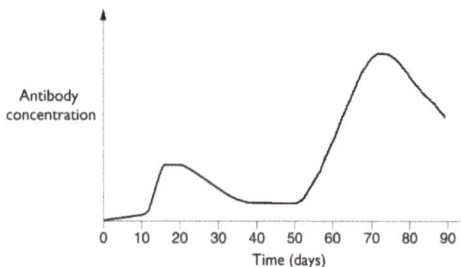

b) A non-competitive inhibitor affects an enzyme by binding to a site other than the active site, known as an allosteric site. This binding alters the enzyme's overall shape, reducing its ability to catalyse the reaction, regardless of whether the substrate is present or not. As a result, the enzyme's activity is decreased because the inhibitor prevents the enzyme from functioning effectively.

## Question 2
A catalyst is a substance that speeds up a chemical reaction without being consumed or permanently altered in the process. It lowers the activation energy required for the reaction to occur, allowing the reaction to proceed more efficiently.

## Question 3
The tertiary structure of an enzyme is crucial because it determines the enzyme's specific three-dimensional shape, which is essential for its function. This shape creates a unique active site where substrates can bind. The precise folding of the enzyme ensures that the active site is correctly aligned to facilitate the chemical reaction. Without this specific structure, the enzyme would not be able to interact effectively with its substrates or catalyse reactions efficiently.

## Question 4
A significant increase in temperature above the enzyme's optimum temperature can lead to a decrease in the rate of an enzyme-controlled reaction. As the temperature rises, the enzyme begins to denature, meaning that the enzyme's active site loses its specific shape. This structural change reduces the enzyme's ability to bind with substrates effectively. Consequently, the reaction rate decreases because fewer enzyme-substrate complexes are formed, and the enzyme's catalytic activity is impaired.

## Question 5

| Macromolecule | Enzyme involved | Monomers |
| --- | --- | --- |
| Starch | Amylase | Maltose |
| Proteins | Protease | Amino acids |
| Lipids | Lipase | Glycerol and three fatty acids |

## Section 2: Enzymes

### Question 6
a) They act as biological catalysts that accelerate biochemical reactions. Without enzymes, these reactions would not happen or occur too slowly to sustain life.

b) The active site of an enzyme is a specific region on the enzyme's surface where substrates bind. It has a unique shape that is complementary to the shape of the substrate molecules. The function of the active site is to facilitate the chemical reaction by positioning the substrates in the correct orientation and lowering the activation energy needed for the reaction to occur.

### Question 7
Competitive inhibitors affect enzyme activity by binding to the active site of the enzyme, competing with the substrate for this binding spot. When a competitive inhibitor is bound to the active site, it prevents the substrate from attaching. This reduces the rate at which the enzyme can catalyse the reaction because fewer enzyme-substrate complexes are formed. However, the inhibition can be overcome by increasing the concentration of the substrate.

### Question 8
Graph B

### Question 9
a) The structure of invertase is crucial for its function because it has a specific active site that is a complementary shape to sucrose. When sucrose binds to the active site, the enzyme lowers the activation energy needed for the reaction, breaking sucrose into glucose and fructose. Any change to the enzyme's structure, such as from high temperatures or pH changes, can alter the active site, stopping it from working.

b) Invertase is active between pH 2 and pH 12, with enzyme activity increasing between pH 2 and 6, and decreasing between 6 and 12. The peak is at pH 6. There is a steeper increase in activity between pH 5 and 6, and a steeper decrease in activity between pH 9 and 10. The minimum activity, at 10%, is observed at pH 12. The enzyme exhibits more activity in acidic conditions (low pH) compared to alkaline conditions (high pH). Explanation: Changes in pH affect the shape of invertase and its active site. At pH 6, the most enzyme-substrate complexes form, resulting in maximum activity. At low or high extremes of pH, the enzyme is partially denatured. At these extreme pH levels, substrate molecules can no longer bind effectively with the enzyme, leading to a decrease in activity.

### Question 10
The lock and key model suggests that the enzyme's active site is exactly shaped to fit a specific substrate, similar to how a key fits into a lock. In this model, the substrate binds to the enzyme without altering the enzyme's structure, as the active site is rigid and does not change shape upon substrate binding. In contrast, the induced fit model proposes that the enzyme's active site is flexible and adjusts its shape to fit the substrate more effectively upon binding. Here, the enzyme undergoes a conformational change to better accommodate the substrate, enhancing the fit and optimizing the interaction for catalysis. This model allows for a more dynamic enzyme shape that adapts to the substrate, leading to improved binding and activity.

# Unit 1: Topic 2: Exchange of nutrients and wastes

## Question 11

Pectinase activity is highest at pH 4 and 35°C: The enzyme exhibits its peak reaction rate of 45 arbitrary units at this combination of pH and temperature, indicating optimal conditions for pectinase activity.
Extreme pH levels and high temperatures reduce pectinase activity: At both pH 2 and pH 8, and at temperatures of 70°C, the enzyme activity drops to 0, suggesting that extreme pH levels and high temperatures can denature the enzyme or inhibit its activity.

## Question 12

At low temperatures, enzyme activity decreases because the enzyme and substrate molecules move more slowly, leading to fewer collisions and reduced formation of enzyme-substrate complexes. As the temperature increases, the enzyme's activity generally rises due to increased molecular movement, which facilitates more frequent collisions between the enzyme and substrate.

However, if the temperature exceeds the enzyme's optimum, the enzyme's structure can become disrupted. High temperatures can cause the enzyme's protein structure to denature, leading to a loss of its specific shape and, consequently, its ability to bind to the substrate effectively. This denaturation reduces or completely stops enzyme activity, as the enzyme can no longer perform its catalytic function.

# Section 3: Exchange Surfaces

### Question 1
To increase the surface area of the small intestine to enhance nutrient absorption.

### Question 2
Villi are tiny, finger-like projections that extend into the lumen of the small intestine, significantly increasing the surface area available for nutrient absorption. Each villus contains a network of capillaries and lymph vessels, which allow absorbed nutrients to be quickly transported into the bloodstream and lymphatic system. Additionally, the epithelial cells on the surface of the villi have microvilli, further increasing the surface area and creating a brush border that enhances nutrient uptake. This extensive surface area and close proximity to the blood and lymphatic vessels ensure that nutrients are efficiently absorbed and transported to the rest of the body.

### Question 3
Keeping blood within vessels maintains high blood pressure, leading to faster blood flow to tissues. It also allows blood to be directed to specific tissues, regulating oxygen and nutrient delivery where needed.

### Question 4
Capillaries have thin walls consisting of a single layer of endothelial cells, which allow for the efficient exchange of gases, nutrients, and waste between the blood and surrounding tissues. Their small diameter and extensive branching provide a large surface area and close proximity to cells, facilitating rapid diffusion. The narrow lumen slows blood flow, enhancing the time available for exchange processes.

### Question 5
a) X: Microvilli; Y: Lacteal; Z: Capillaries

b) Active transport differs from diffusion in that it requires energy, usually in the form of ATP, to move substances against their concentration gradient, from an area of lower concentration to higher concentration. In contrast, diffusion is a passive process that moves substances down their concentration gradient, from an area of higher concentration to lower concentration, without the use of energy.

c) Microvilli significantly increase the surface area of the epithelial cells in the villi, enhancing their ability to absorb nutrients efficiently.

### Question 6
A closed circulatory system is one in which blood is contained within vessels, such as arteries, veins, and capillaries, and is pumped by the heart throughout the body. This system keeps the blood separate from the body's other fluids, allowing for efficient transport of oxygen, nutrients, and waste products to and from tissues.

# Unit 1: Topic 2: Exchange of nutrients and wastes

## Question 7
Children with untreated coeliac disease experience slower growth and increased fatigue because their immune response to gluten damages the villi in their small intestine, causing them to become blunted or atrophic. The villi are responsible for absorbing nutrients such as carbohydrates, proteins, fats, vitamins, and minerals. When the villi are damaged and flattened, their surface area is significantly reduced, impairing nutrient absorption. This means that children with untreated coeliac disease are unable to efficiently absorb the essential nutrients needed for growth and energy production, leading to slower growth rates and increased fatigue due to nutritional deficiencies.

## Question 8
a) Lack of cell wall, chloroplasts or large permanent vacuole. Presence of microvilli.

b) A: Plasma membrane; B: Nucleus; C: Nucleolus; D: Microvilli

# Section 4: Excretory System

## Question 1
a) Nephron

b) S: Renal artery; T: Glomerulus; U: Bowman's Capsule; V: Proximal convoluted tubule; W: Distal convoluted tubule; X: Collecting duct; Y: Loop of Henle; Z: Renal vein

c) Osmoregulation

d) Needs to point to point V (proximal convoluted tubule)

e) In desert-dwelling mammals, the loop of Henle would likely be much longer compared to those in mammals living in less arid environments. The longer loop of Henle allows for a greater reabsorption of water from the filtrate back into the bloodstream, creating highly concentrated urine. This adaptation is crucial for conserving water in desert conditions, where water is scarce and dehydration is a significant risk. By extending the loop, these animals can efficiently conserve water, reducing the need for frequent drinking and enhancing their survival in arid environments.

f) Glomerular filtration, or ultrafiltration, occurs in the kidney when blood enters the glomerulus, a network of capillaries within the Bowman's capsule. High blood pressure forces water, urea, ions, glucose (small molecules) out of the blood through the capillary walls, while larger molecules like proteins and blood cells are prevented from filtering. The filtered fluid, known as filtrate, enters the Bowman's capsule and moves into the renal tubule for further processing.

## Question 2
As the filtrate passes through the proximal convoluted tubule, water, glucose, amino acids, and ions are reabsorbed back into the bloodstream, leaving urea behind, which increases its concentration. In the loop of Henle, further water reabsorption occurs, especially in the descending limb, while the ascending limb actively reabsorbs salts without water, creating a concentrated medullary gradient. In the distal convoluted tubule and collecting duct, more water is reabsorbed under the influence of antidiuretic hormone (ADH), further concentrating urea. This selective reabsorption process significantly reduces the volume of water in the filtrate, leading to urine with a much higher concentration of urea compared to the original glomerular filtrate.

## Question 3
a) V: Afferent arteriole; W: Efferent arteriole; X: Glomerulus; Y: Bowman's Capsule; Z: Proximal convoluted tubule

b) Glucose, water

c) Glomerular filtration (ultrafiltration)

d) The afferent arteriole, which brings blood into the glomerulus, has a larger diameter compared to the efferent arteriole, which carries blood away from the glomerulus. This difference in diameter creates higher pressure within the glomerulus. The increased pressure is essential for ultrafiltration, as it forces water and small solutes from the blood through the filtration membrane into the Bowman's capsule, forming the glomerular filtrate.

# Unit 1: Topic 2: Exchange of nutrients and wastes

## Question 4

Under healthy conditions, the glomerular filtration barrier allows only small molecules and fluids to pass into the filtrate, while larger proteins like albumin are retained in the bloodstream. When the kidneys are damaged, the filtration barrier can become compromised, leading to an increased permeability. This damage allows larger proteins, such as albumin, to leak into the urine.

# TOPIC 3: CELLULAR ENERGY, GAS EXCHANGE AND PLANT PHYSIOLOGY

## Section 1: Respiration

### Question 1
$C_6H_{12}O_6 + 6O_2 \rightarrow 6CO_2 + 6H_2O + 36\text{-}38 \text{ ATP}$

### Question 2
Catabolism refers to the breakdown of complex molecules into simpler ones. This process releases energy, which is often captured in the form of ATP. Anabolism involves the synthesis of complex molecules from simpler ones. This process requires an input of energy.

### Question 3
Catabolic reaction: Glycolysis; Anabolic reaction: Protein synthesis, DNA replication, Photosynthesis.

### Question 4
Aerobic respiration produces a larger amount of ATP per glucose molecule. Aerobic respiration yields approximately 36-38 ATP molecules per glucose, while anaerobic respiration typically produces only 2 ATP molecules per glucose.

### Question 5
Oxygen; Glucose

### Question 6
a) X: Glucose; Y: ATP; Z: Krebs Cycle
b) Cristae of the mitochondria

### Question 7

| Feature | Aerobic Respiration | Anaerobic Respiration |
|---|---|---|
| Oxygen requirement | Requires oxygen | Does not require oxygen |
| Location in cells | Mitochondria | Cytoplasm |
| Energy yield | High | Low |
| End products | Carbon dioxide and water | Lactic acid (in animals) or ethanol and carbon dioxide (in yeast) |
| Efficiency | High | Low |
| ATP production | 36-38 ATP per glucose | 2 per glucose molecule |
| Process stages | Glycolysis, Krebs cycle, Electron Transport Chain | Glycolysis only |

# Unit 1: Topic 3: Cellular energy, gas exchange and plant physiology

## Question 8
ATP acts as an energy carrier in cells. During catabolic reactions energy is released and stored in ATP molecules. This energy is then used to drive anabolic reactions, which build complex molecules from simpler ones. The ATP provides the necessary energy by transferring a phosphate group through hydrolysis, releasing energy that is used to power various cellular processes.

## Question 9

|  | Glycolysis | Krebs Cycle | Electron transport chain |
|---|---|---|---|
| **Location in cell** | Cytoplasm | Mitochondrial matrix | Inner mitochondrial membrane (cristae) |
| **Net inputs** | 1 x glucose | 2 x acetyl CoA | 10 x NADH<br>2 x FADH |
| **Net outputs** | 2 x pyruvate<br>2 x ATP<br>2 NADH | 2 x ATP<br>6 x NADH<br>2 x FADH<br>4 x $CO_2$ | TP |

## Question 10
a) Glucose + Oxygen → Carbon dioxide + Water + ATP

b) Glucose → Ethanol + Carbon dioxide

c) Lactic acid is produced, instead of ethanol

d) During intense exercise.

# Section 2: Photosynthesis

## Question 1

a) $6CO_2 + 6H_2O \rightarrow C_6H_{12}O_6 + 6O_2$

b) ATP, NADPH

c)
  i) Thylakoid (granum is the stack of thylakoids)
  
  ii)

| | |
|---|---|
| Name of the stage of photosynthesis occurring at X | Light dependent stage |
| Input molecules required at X | Water, Light (photons), NADP, ADP |
| Output molecules produced at X | Oxygen, ATP, NADPH |

## Question 2

a) Light independent stage / Calvin cycle

b) ATP: provides energy to drive reactions to create GP; NADPH: provide electrons and hydrogen to convert carbon dioxide into sugars.

c) The transfer of electrons helps to generate a proton gradient across the membrane, which is used to produce ATP and NADPH, needed for the Calvin cycle to synthesise sugars.

d)

| | Light-dependent stage | Light-independent stage |
|---|---|---|
| Location in cell | Thylakoid membrane of the chloroplast | Stroma of the chloroplast |
| Net inputs | Water, light energy (photons), NADP, ADP | Carbon dioxide, ATP, NADPH |
| Net outputs | Oxygen, ATP, NADPH | Glucose, ADP, NADP |

# Section 3: Gas Exchange

### Question 1
a) Increases the surface area available for gas exchange, allowing for more efficient uptake of oxygen and removal of carbon dioxide.

b) Reduces the distance that gases must diffuse, facilitating quicker and more efficient gas exchange between the air and the blood.

c) Ensures close proximity between the blood and the air in the alveoli, enhancing the efficiency of gas exchange by maximising the contact area for oxygen and carbon dioxide diffusion.

### Question 2
Oxygen will diffuse from the area of higher partial pressure in the alveoli to the area of lower partial pressure in the pulmonary capillaries.

### Question 3
a) Presence of nucleus.

b) Rough endoplasmic reticulum and Golgi apparatus. The RER is involved in the production of proteins, while the Golgi apparatus processes and packaging these surfactant proteins into vesicles for secretion.

c) Thin walls, moist, large surface area, good blood supply

d) Active transport: Molecules or ions move against their concentration gradient. This process requires energy, typically in the form of ATP, to pump substances through specific transport proteins in the membrane.
Facilitated Diffusion: Molecules or ions move down their concentration gradient with the help of specific transport proteins or channels. This process does not require energy; instead, it relies on these proteins to provide a pathway for substances that cannot directly diffuse through the lipid bilayer.

e) This shape increases the surface area-to-volume ratio, allowing for more efficient diffusion of oxygen and carbon dioxide across the cell membrane. The larger surface area facilitates quicker and more effective gas exchange, ensuring that red blood cells can deliver oxygen to tissues and remove carbon dioxide more efficiently than if they were spherical.

# Section 4: Plant Physiology

## Question 1
The loss of water vapour from the leaves of a plant.

## Question 2
Xylem is described as a tissue because it is a group of cells that work together to perform a specific function. Xylem tissue is composed of various cell types, including tracheids, vessel elements, and fibres and it is responsible for the transport of water and dissolved minerals from the roots to the rest of the plant, and provides structural support.

## Question 3

|  | Xylem | Phloem |
|---|---|---|
| Function | Transports water and dissolved minerals from roots to other parts of the plant. | Transports organic nutrients, especially sugars, from leaves to other parts of the plant. |
| Cell types | Includes vessel elements, tracheids, xylem parenchyma and xylem fibres. | Includes sieve tube elements, companion cells, phloem parenchyma, and phloem fibres. |
| Direction of flow | Unidirectional (from roots to shoots | Bidirectional |
| Living/dead cells | Contains both living (xylem parenchyma) and dead cells (vessels and tracheids). | Contains living cells (sieve tube elements and companion cells) but also has some dead cells (phloem fibres). |

## Question 4

| Condition | Effect on transpiration rate | Explanation |
|---|---|---|
| Increased temperature | Increases | Higher temperatures increase the rate of water evaporation from the leaf surfaces, leading to an increased concentration gradient. |
| Increased wind speed | Increases | Increased wind speed removes water vapour from around the leaf surface more quickly, increasing the concentration gradient |
| Increased humidity | Decreases | Higher humidity reduces the concentration gradient between the inside of the leaf and the external environment, slowing the rate of water evaporation. |
| Decreased light intensity | Decreases | Lower light intensity reduces the rate of photosynthesis, which decreases the production of sugars and lowers the stomatal opening. |

### Question 5

Light triggers the activation of guard cells causing them to take up potassium ions, which increases their osmotic pressure and causes water to flow in, making the cells turgid. As the guard cells swell, they change shape, causing the stomatal pores to open.

### Question 6

Light Intensity: Transpiration rate increases with light intensity. At low light intensity (500 lux), the transpiration rate is 0.15 g/hr, while at high light intensity (2000 lux), it rises to 0.5 g/hr. This suggests that higher light intensity promotes more stomatal opening and enhances transpiration.

Temperature: Transpiration rate also increases with temperature. At low temperature (10°C), the rate is 0.1 g/hr, whereas at high temperature (30°C), it is 0.55 g/hr. This indicates that higher temperatures increase the rate of water evaporation from leaf surfaces, leading to higher transpiration.

Combined Effect: When comparing the control condition with varying light intensities and temperatures, the highest transpiration rate is observed at high light intensity and high temperature (0.55 g/hr), showing that both high light intensity and high temperature together maximise the rate of transpiration.

# Section 4: Plant Physiology

# Unit 2

# Maintaining the internal environment

# TOPIC 1: HOMEOSTASIS - THERMOREGULATION AND OSMOREGULATION

## Section 1: Nervous System

### Question 1
When a nerve impulse arrives at a synapse, it causes depolarisation of the presynaptic membrane. This depolarisation opens voltage-gated calcium channels, allowing calcium ions ($Ca^{2+}$) to flow into the presynaptic terminal. The influx of calcium ions triggers the fusion of neurotransmitter-containing vesicles with the presynaptic membrane. As a result, the neurotransmitters are released into the synaptic cleft through exocytosis, where they can then bind to receptors on the postsynaptic membrane, continuing the transmission of the nerve impulse.

### Question 2
Stimuli; Receptors; Stimulus

### Question 3
By sodium-potassium pumps actively transporting three sodium ions ($Na^+$) out of the neuron and two potassium ions ($K^+$) into the neuron, creating a negative charge inside the cell relative to the outside.

### Question 4
a) Neuron A: Sensory/Afferent; Neuron B: Interneuron; Neuron C: Motor/Efferent; X: Receptor; Y: Effector

b) Arrow should be positioned at the gap between Neuron A and B, or Neuron B and C.

### Question 5
a) Sensory/Afferent

b) In sensory neurons, the soma is typically positioned in the middle of the neuron. In motor neurons and interneurons, the soma is typically positioned at one end of the neuron.

### Question 6
Synpase

## Section 1: Nervous System

### Question 7

| Name | Letter | Function |
|---|---|---|
| Receptor | V | Detects specific stimuli from the environment and converts them into electrical signals to be sent to the sensory neurons. |
| Sensory neuron | W | Transmits electrical signals from receptors to the central nervous system (CNS), where the information is processed. |
| Interneuron | X | Connects sensory neurons to motor neurons within the CNS, facilitating communication and integration of information. |
| Motor neuron | Y | Carries electrical signals from the CNS to effectors, such as muscles or glands, to produce a response. |
| Effector | Z | Executes the response to a stimulus, such as a muscle contracting or a gland secreting hormones. |

### Question 8
Homeostasis

### Question 9
a) -70 mV

b) X: Depolarisation; Y: Repolarisation

c) X: The neuron's membrane potential becomes more positive. This occurs when voltage-gated sodium channels open in response to a stimulus, allowing sodium ions (Na⁺) to rush into the neuron. The influx of positively charged sodium ions reduces the negative charge inside the cell, leading to a rapid increase in membrane potential and the generation of an action potential.

Y: The neuron's membrane potential returns to its resting negative value after depolarisation. This happens when voltage-gated sodium channels close and voltage-gated potassium channels open, allowing potassium ions (K⁺) to exit the cell. The efflux of positively charged potassium ions restores the negative internal charge of the neuron, bringing the membrane potential back to its resting state.

### Question 10
-70 mV; depolarisation; threshold potential

### Question 11

| Sensory neurons | Motor neurons |
|---|---|
| Soma in centre of cell | Soma at side of cell |
| Long dendrite | Short dendrites |
| Short axon | Long axon |

## Unit 2: Topic 1: Homeostasis - thermoregulation and osmoregulation

### Question 12

| Sensory receptor | Stimulus detected |
|---|---|
| Photoreceptor | Light |
| Mechanoreceptor | Mechanical pressure, vibration, touch |
| Thermoreceptor | Temperature |
| Nociceptor | Pain |
| Chemoreceptor | Chemicals |
| Baroreceptor | Pressure changes |

### Question 13
Tetrodotoxin affects the initiation and propagation of action potentials by blocking voltage-gated sodium channels on the postsynaptic membrane. When tetrodotoxin is present, it binds to and inhibits these sodium channels, preventing them from opening in response to neurotransmitter binding. This blockage stops the influx of sodium ions into the postsynaptic neuron, which is necessary for depolarisation and the generation of an action potential.

### Question 14
Myelin acts as an insulator. This insulation allows the action potential to jump between nodes of Ranvier in a process called saltatory conduction. This speeds up impulse transmission compared to continuous conduction in a non-myelinated axon, where the impulse travels along the entire membrane.

# Section 2: Endocrine System

### Question 1
Homeostasis is the process by which an organism maintains a stable internal environment, such as temperature, pH, and fluid balance, despite changes in external conditions.

### Question 2
The difference in response to the hormone between the two groups of cells is due to the presence of specific receptors. The group of cells that responded to the hormone had receptors on their surfaces or inside the cell that were specific to that hormone, allowing it to bind and trigger a response. The neighbouring group did not respond because they lacked the necessary receptors, preventing the hormone from binding and initiating any cellular response.

### Question 3
Negative feedback mechanisms maintain homeostasis by reversing changes in the body. When a change is detected, receptors send signals to a control centre, like the brain. The control centre then activates effectors, such as glands or muscles, to correct the change and restore balance. Once normal conditions are reached, the response stops, completing the feedback loop.

### Question 4
When blood glucose levels are high, the pancreas releases insulin, a hormone that helps lower blood glucose. Insulin prompts cells, especially in the liver and muscles, to take in glucose from the blood and store it as glycogen.

### Question 5
Stimulus → Receptor → Control Centre → Effector → Response

### Question 6
a)  Endocrine cells affect specific targets because their hormones bind only to cells with matching receptors.
b)  A tissue may become less responsive to a hormone due to receptor downregulation. This happens when cells reduce the number of receptors on their surface or the receptors become less sensitive to the hormone, often due to prolonged or excessive exposure, making the tissue less responsive.

### Question 7
The normal rats maintained their blood glucose levels by converting stored glycogen in the liver into glucose through the action of glucagon. This glucose is then released into the bloodstream to keep blood glucose levels stable.

### Question 8
Stimulus: increased demand for oxygen; Receptor: Baroreceptors in aorta and carotid arteries; Control centre: Medulla oblongata; Effector: Heart; Response: Increased heart rate to supply more oxygen.

# Unit 2: Topic 1: Homeostasis - thermoregulation and osmoregulation

## Section 3: Thermoregulation

### Question 1
Nervous Mechanisms:
Sweating: Temperature receptors in the skin and hypothalamus detect the rise in temperature. The nervous system responds by sending signals to the sweat glands to increase perspiration. The evaporation of sweat from the skin surface helps cool the body by dissipating excess heat.
Vasodilation: The nervous system also causes the blood vessels near the skin's surface to dilate, or widen. This process, known as vasodilation, increases blood flow to the skin, where heat can be lost to the environment more effectively.

Endocrine Mechanisms
Antidiuretic Hormone (ADH): The pituitary gland releases ADH to help control fluid balance. When the body is overheating, ADH levels can be adjusted to regulate water retention and ensure proper hydration, which supports efficient sweating and cooling.
Thyroid Hormones: The endocrine system adjusts thyroid hormone levels in response to elevated body temperature. The thyroid gland reduces the secretion of thyroid hormones, which are involved in increasing metabolic rate and generating heat. By lowering the levels of these hormones, the body decreases its heat production, helping to cool down.

### Question 2
An organism that regulates its body temperature internally through metabolic processes, maintaining a relatively constant temperature regardless of external environmental conditions.

### Question 3
Evaporative cooling helps organisms regulate their body temperature by using the heat from their bodies to convert a liquid into a gas. When a liquid, such as sweat or water, evaporates from the surface of the skin or other body parts, it absorbs heat from the surrounding environment. This process removes heat from the body, resulting in a cooling effect. Factors like a larger surface area, increased airflow, and lower humidity can enhance the rate of evaporation, making this cooling mechanism more effective. By dissipating excess heat through the evaporation of liquids, evaporative cooling helps maintain a stable internal temperature.

### Question 4
Structural: Thick layer of blubber and thick water-repellent fur that provides insulation.
Physiological: Vasoconstriction, where blood vessels constrict to reduce blood flow to the extremities, minimising heat loss from the surface of the skin; high metabolic rate which produces a lot of body heat, brown adipose tissue for thermogenesis, have specialised countercurrent heat exchange systems, where arteries and veins run closely together, allowing heat from warm arterial blood flowing to the extremities to be transferred to the cooler venous blood returning to the core.

# Section 3: Thermoregulation

## Question 5
When the body needs to cool down, blood vessels in the skin dilate, or widen, in a process known as vasodilation. This increased blood flow to the skin's surface enhances the transfer of heat from the body to the environment. As the warm blood moves closer to the skin's surface, more heat is released into the surrounding. when the body needs to conserve heat, blood vessels in the skin constrict, or narrow, in a process known as vasoconstriction. This reduces the volume of blood flowing to the skin, minimising heat loss by limiting the amount of warm blood reaching the skin's surface.

## Question 6
Aestivation: A period of dormancy or inactivity that some animals enter during hot or dry conditions to conserve energy and avoid adverse environmental conditions.
Torpor: A state of reduced metabolic rate that allows animals to conserve energy during periods of reduced food availability or adverse conditions.
Kleptothermy: Where an animal relies on the body heat of another animal or its environment to regulate its own body temperature.
Hibernation: A prolonged state of inactivity and reduced metabolic rate that animals enter during cold periods to conserve energy and survive through winter when food is scarce.

## Question 7
White adipose tissue: acts as an effective insulator. This layer of fat helps to reduce heat loss by providing a barrier against the cold environment. It stores energy that can be mobilised when food is scarce, which is particularly useful during long periods of extreme cold and reduced food availability.
Brown adipose tissue: specialised for thermogenesis, which is the production of heat. The high density of mitochondria in brown fat cells enables rapid heat generation through a process called non-shivering thermogenesis. This process helps polar bears maintain their core body temperature by generating heat

# Unit 2: Topic 1: Homeostasis - thermoregulation and osmoregulation

## Section 4: Osmoregulation

### Question 1
Guard cells

### Question 2
When ABA binds to receptors on guard cells, it activates ion channels that cause potassium and other ions to leave the cells. This loss of ions decreases the osmotic pressure inside the guard cells, leading to water exiting the cells through osmosis. As the guard cells lose turgor pressure, they become flaccid and close the stomatal pores. This process helps the plant conserve water.

### Question 3
a) Guard cells

b) Transpiration

c) Increasing humidity, decreasing temperature, reducing wind speed, reducing light intensity.

### Question 4
a) The cuticle reduces water loss by acting as a protective, waxy layer on the surface of plant leaves and stems. This waxy coating is hydrophobic, meaning it repels water. By creating a barrier that prevents water from evaporating directly through the leaf surface, the cuticle significantly reduces the rate of water loss from the plant.

b) Guard cells regulate the size of the stomata, reducing water loss by closing the pores in response to environmental conditions; Leaf hairs trap humid air near the leaf surface, reducing the concentration gradient and lowering the rate of water vapour loss.

### Question 5
a) Hypothalamus

b) Pituitary gland

c) When ADH is released, it increases the permeability of the kidney's collecting ducts to water. This allows more water to be reabsorbed back into the bloodstream, resulting in more concentrated urine and reduced urine volume.

### Question 6
a) Xerophytes

b) Plants lose water to the environment primarily through transpiration, when water vapour exits the plant through stomata on the leaf surface.

# Section 4: Osmoregulation

## Question 7

Detection: Osmoreceptors in the hypothalamus sense changes in blood osmolarity. If blood becomes too concentrated, indicating low water levels, the osmoreceptors detect this imbalance.

Response: The hypothalamus responds by stimulating the pituitary gland to release antidiuretic hormone (ADH).

Adjustment: ADH acts on the kidneys to increase water reabsorption, reducing blood osmolarity and balancing water levels in the body.

Feedback: When blood osmolarity returns to normal, the reduced stimulus causes the hypothalamus to decrease ADH secretion, maintaining equilibrium.

## Question 8

a)

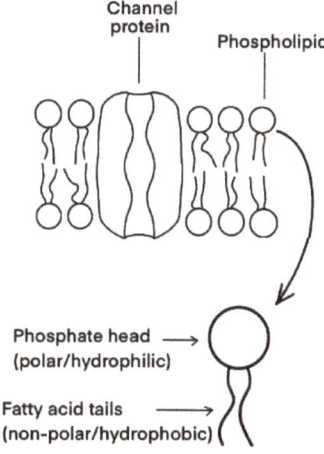

b) Ions cannot easily pass through the lipid bilayer of the cell membrane due to their charge and size. Channel proteins allow ions to move across the membrane by facilitated diffusion, which requires no energy and helps maintain proper ion balance within the cell.

c) Without the outward movement of ions, guard cells cannot lose turgor pressure, preventing stomata from closing. This means the plant cannot regulate water loss effectively, leading to increased transpiration.

## Question 9

Water Lily Adaptations: Water lilies have stomata only on the upper epidermis because their lower surface is submerged in water. This arrangement facilitates gas exchange with the air while reducing water loss.

Eucalypt Adaptations: Eucalypts have more stomata on their lower epidermis, which is shaded from direct sunlight. This helps them conserve water in their dry, terrestrial environment.

Water Regulation: Water lilies do not need many stomata on the lower side due to their aquatic habitat, while eucalypts use a higher number of stomata on the lower side to manage water loss effectively.

Gas Exchange Needs: Water lilies primarily need gas exchange with the air above, while eucalypts balance gas exchange with water conservation, leading to their specific stomatal distribution.

# TOPIC 2: INFECTIOUS DISEASE AND EPIDEMIOLOGY

## Section 1: Disease

### Question 1
Infectious diseases are caused by pathogens and can be spread between people, while non-infectious diseases result from genetic, environmental, or lifestyle factors.

### Question 2

| Pathogen | Cellular or non-cellular | Eukaryotic or prokaryotic |
|---|---|---|
| Prion | Non-cellular | |
| Virus | Non-cellular | |
| Bacterium | Cellular | Prokaryotic |
| Fungus | Cellular | Eukaryotic |
| Protist | Cellular | Eukaryotic |

### Question 3
Adhesion: Pathogens often have specialised structures, such as pili or fimbriae, that enable them to attach firmly to the host's tissues.

Capsules: Many pathogens have a protective capsule made of polysaccharides or proteins surrounding their cell wall. This capsule helps the pathogen evade the host's immune system by making it harder for immune cells to recognise and engulf the pathogen.

Toxins: Pathogens can produce toxins, which are harmful substances that damage host tissues or disrupt normal cellular functions. Toxins can lead to disease symptoms and enhance the pathogen's ability to establish an infection.

Invasion Factors: Some pathogens produce enzymes or proteins called invasion factors that help them penetrate host tissues. These factors break down host cell barriers, allowing the pathogen to invade deeper into tissues and spread more easily.

### Question 4
Endotoxins are part of the outer membrane of (Gram-negative) bacteria and are released when these bacteria die. In contrast, exotoxins are secreted by bacteria during their growth and metabolism.

# Section 1: Disease

## Question 5
An infectious agent that causes disease.

## Question 6

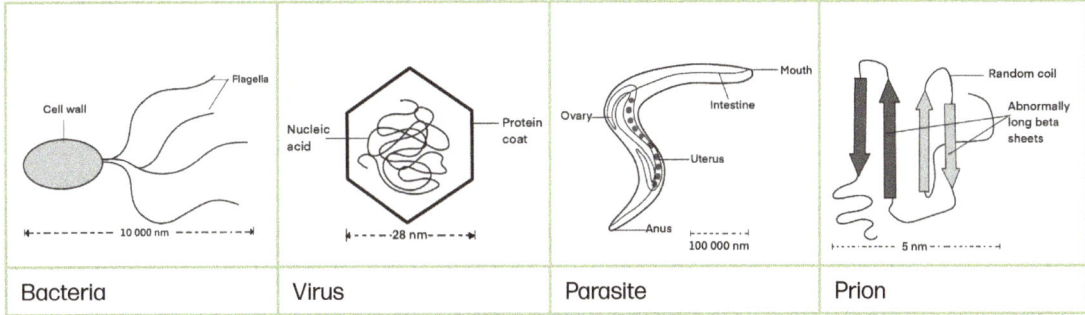

| Bacteria | Virus | Parasite | Prion |

## Question 7
Thick slime capsule: The capsule helps Streptococcus pneumoniae evade the immune system by making it harder for phagocytes to engulf the bacteria and preventing immune recognition.
Pili: Pili allow the bacterium to adhere to host tissues, aiding in colonization and infection by resisting removal through coughing or mucous flow.

## Question 8
Adherence factor: Pili and fimbriae; Invasion factor: Collagenase and hyaluronidase; Toxin: Botulinum neurotoxins (BoNTs)

# Unit 2: Topic 2: Infectious disease and epidemiology

## Section 2: Immune Response

### Question 1
a) An antigen is a foreign substance that triggers an immune response, typically a protein or polysaccharide on the surface of pathogens. An antibody is a protein produced by the immune system that specifically binds to an antigen, helping to neutralize or eliminate the pathogen.

b) When the same antigen enters the body for a second time, antibody production is faster and higher due to the presence of memory cells. These cells, formed during the first exposure, quickly recognize the antigen and trigger a swift and robust immune response.

### Question 2
Active immunity: Babies can develop active immunity to chickenpox through vaccination. The vaccine stimulates their immune system to produce antibodies and memory cells against the virus.

Passive immunity: Babies can receive passive immunity from their mothers through the placenta during pregnancy. This provides temporary protection as maternal antibodies are transferred to the baby, helping to protect them from chickenpox in the early months.

### Question 3

| Plant defence | Type of defence | |
| --- | --- | --- |
| | Chemical | Physical |
| Thick, waxy cuticle | | ✓ |
| Plant defensins | ✓ | |
| Poisonous berries | ✓ | |
| Bark that falls off | | ✓ |

### Question 4
The humoral immune response involves B cells producing antibodies that circulate in the blood. These antibodies bind to specific antigens on pathogens, marking them for destruction or neutralisation. In contrast, the cell-mediated immune response relies on T cells that directly attack infected or cancerous cells. Instead of producing antibodies, T cells identify and destroy cells displaying specific antigens on their surface. The humoral response is effective against extracellular pathogens like bacteria, while the cell-mediated response targets intracellular pathogens such as viruses and cancer cells.

### Question 5
Phagocytes such as neutrophils and macrophages engulf and digest pathogens. Natural killer cells are responsible for identifying and destroying infected or cancerous cells.

## Section 2: Immune Response

### Question 6
Immune system cells distinguish between self and non-self by recognising specific markers on cell surfaces. Self cells have unique molecules called Major Histocompatibility Complex (MHC) molecules that signal they belong to the body. Non-self cells, such as pathogens, lack these MHC markers or present foreign antigens, triggering an immune response.

### Question 7
a) Viruses are classified as non-cellular pathogens. They do not have cellular structures or functions, such as a cell membrane or organelles. Instead, they consist of genetic material enclosed in a protein coat and rely on host cells to replicate.

b) The body's first line of defence against the dengue virus includes: Physical barriers: The skin acts as a physical barrier, preventing the virus from entering the body through cuts or abrasions; Chemical barriers: Secretions such as saliva and mucus contain antimicrobial substances that can inhibit or destroy pathogens, including viruses.

### Question 8
Vaccination protects a person from disease by stimulating the immune system to recognize and respond to specific pathogens. When a vaccine is administered, it introduces a harmless part of the pathogen, such as a protein or an inactivated form of the virus, into the body. This exposure prompts the immune system to produce a primary immune response, including the production of antibodies and memory cells specific to the pathogen. If the person is later exposed to the actual pathogen, their immune system can quickly recognize and attack it, preventing illness. This provides immunity without causing the disease.

### Question 9
Complement proteins circulate in the blood and, upon activation, help to destroy pathogens. They work through a process called the complement cascade, which enhances the ability of antibodies and phagocytes to clear microbes and damaged cells. Complement proteins can directly lyse pathogens by forming a membrane attack complex that creates pores in the pathogen's membrane, leading to its destruction.

Natural Killer (NK) cells target and destroy infected or cancerous cells. They detect cells that lack "self" markers, such as those altered by viral infections, and induce apoptosis (programmed cell death) in these compromised cells, preventing the spread of infection.

### Question 10
Blood vessels dilate during inflammation, which boosts blood flow to the affected area, delivering more immune cells, oxygen, and nutrients necessary for healing while aiding in the removal of waste products. Additionally, increased permeability of blood vessel walls allows immune cells and proteins to more easily move from the bloodstream into the tissues, enabling the immune system to more effectively target and neutralise pathogens.

### Question 11
A foreign substance (usually protein) that can provoke an immune response when detected by the body.

# Unit 2: Topic 2: Infectious disease and epidemiology

## Question 12
Streptomycin was the most effective antibiotic, as evidenced by its largest zone of inhibition, indicating the strongest antibacterial activity among the tested antibiotics. Tetracycline and amoxicillin showed comparable effectiveness, though both were less effective than streptomycin. Penicillin exhibited only a minor effect on bacterial growth. In contrast, neomycin and sulfonamide had no measurable impact on the bacteria, suggesting they were ineffective in this experiment.

## Question 13
During inflammation, blood vessels dilate, increasing blood flow to the affected area. This increased blood flow causes the area to appear red. Second, the permeability of the blood vessel walls increases, allowing immune cells, fluid, and proteins to move from the blood into the surrounding tissue. This fluid accumulation leads to swelling.

## Question 14
a) Glycosidic

b) Water

c) Lysozyme cannot destroy viruses because viruses do not have cell walls, which are the target of lysozyme. Instead, viruses consist of genetic material enclosed in a protein coat, and lysozyme's action is specific to breaking down bacterial cell walls, not viral structures.

## Question 15
Phagocytes such as neutrophils and macrophages destroy pathogens through a process called phagocytosis. During phagocytosis, the phagocyte engulfs the pathogen. Lysosomes then release, enzymes and toxic substances to break down the pathogen. The pathogen's remnants are then expelled from the phagocyte.

## Question 16
The treatment with rabies immunoglobulin involves injecting antibodies directly into the body. These antibodies are specific to the rabies virus and help neutralize the virus before it can cause an infection. Passive immunity is used in this case because it provides immediate protection by directly supplying the necessary antibodies. This is crucial because rabies progresses rapidly, and there isn't enough time for the body to mount its own immune response through active immunity.

## Question 17
Natural killer (NK) cell.

## Question 18
D → C → A → F → B → E

## Question 19
B lymphocytes (B cells) produce and secrete antibodies that bind to specific antigens on pathogens, leading to their neutralisation or marking for destruction by other immune cells. They also form memory B cells that enable a quicker response upon re-exposure to the same antigen. In contrast, T cells are central to the cell-mediated response and do not produce antibodies. Instead, they directly interact with and eliminate infected cells. Cytotoxic T cells target and kill infected cells, helper T cells release cytokines to activate B cells and other

immune cells, and regulatory T cells help modulate the immune response and prevent autoimmunity. Thus, B cells focus on antibody production and pathogen neutralization, while T cells directly destroy infected cells and coordinate the immune system's activities.

## Question 20
Memory cells formed during gene therapy can hinder long-term treatment by recognising and attacking the viral vector used for the therapy. If the viral vector is encountered again, these memory cells mount a strong immune response, neutralising the vector before it delivers the therapeutic gene. This reduces the effectiveness of the therapy over time.

## Question 21
a) Line should increase after Day 50.

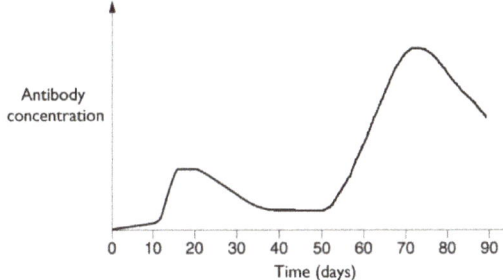

b) After Day 50, there is a rapid and high increase in antibody levels upon subsequent exposure to the same antigen. This is due to the presence of memory cells from the primary response, which quickly recognise the antigen and stimulate a higher antibody response.

# Section 3: Epidemiology

### Question 1

a) Direct Contact: Through physical touch with an infected person or animal, such as shaking hands or hugging; Contact with Body Fluids: By coming into contact with bodily fluids like blood, saliva, or urine from an infected individual; Contaminated Food: By ingesting food contaminated with pathogens, often due to improper handling, cooking, or storage; Contaminated Water: By drinking or coming into contact with water that has been contaminated with pathogens; Disease-Specific Vectors: Through vectors such as insects (e.g., mosquitoes, ticks) that transmit pathogens to humans or animals. Describe how the body's first line of defence protects against pathogens.

b) Physical Barriers: The skin acts as a physical barrier, preventing pathogens from entering the body. Mucous membranes in the respiratory, gastrointestinal, and urogenital tracts also trap and expel pathogens; Chemical Barriers: Secretions like sweat, saliva, and tears contain antimicrobial substances, such as lysozyme, that can destroy pathogens. Additionally, stomach acid provides an acidic environment that kills many ingested microbes.

### Question 2

An infected person should practice good hygiene by regularly washing their hands with soap and water, or using hand sanitiser when soap is not available, and avoid touching their face, particularly the eyes, nose, and mouth. Additionally, they should stay home from work, school, or other public places while symptomatic and avoid close contact with others, such as hugging, kissing, or shaking hands, to reduce the risk of transmission.

### Question 3

Contact tracing involves health authorities identifying individuals diagnosed with an infectious disease and then determining who they have been in close contact with during their infectious period. Those identified contacts are subsequently notified about their potential exposure. They are advised to monitor for symptoms, get tested and follow recommended preventive measures such as self-isolation or quarantine. This approach helps to prevent further spread of the disease by ensuring that those who might be infected are promptly informed and managed.

### Question 4

Herd immunity is achieved when a significant proportion of a population becomes immune to an infectious disease, either through vaccination or previous infection. The key principles include reducing disease transmission because immune individuals act as barriers, limiting the chance of infected persons coming into contact with susceptible individuals. This creates protection for those who are not immune, such as individuals who cannot be vaccinated for medical reasons, because the overall prevalence of the disease in the community is lower.

### Question 5

Quarantine involves isolating individuals who may have been exposed to the disease but are not yet showing symptoms. The process involves separating these individuals from the general population to prevent potential transmission. During quarantine, individuals are monitored for symptoms and provided with medical care if needed. This measure helps to limit the spread of the disease by ensuring that those who might be infectious do not come into contact with others.

## Section 3: Epidemiology

### Question 6
A person with a mild *E. coli* infection should practice good hand hygiene by washing their hands thoroughly with soap and water, especially after using the toilet and before handling or eating food. Additionally, they should avoid preparing food for others until they have fully recovered to reduce the risk of contaminating food and potentially spreading the bacteria.

### Question 7
Regularly cleaning and disinfecting frequently-touched surfaces, such as doorknobs, countertops and light switches, helps remove pathogens. Properly storing and cooking food prevents contamination and the growth of harmful bacteria. Ensuring good ventilation by opening windows and using exhaust fans can reduce the concentration of airborne pathogens. Additionally, maintaining personal hygiene by encouraging frequent handwashing with soap and water, especially before eating or after using the toilet, further reduces the risk of spreading infections.

### Question 8
Increased travel patterns, such as frequent long-distance journeys or international trips, can facilitate the rapid transmission of diseases across regions and countries. High population density, particularly in urban areas, enables pathogens to spread more easily due to close contact between individuals. Social interactions, including gatherings, public events, and communal activities, further contribute to the spread as they create opportunities for pathogens to move from one person to another. Overall, higher mobility, dense populations, and frequent social interactions can accelerate the spread of infectious diseases, making it crucial to implement control measures to manage outbreaks effectively.

### Question 9
The vaccination campaign achieved the highest reduction in cases (70%) compared to the other strategies, indicating it was the most effective in decreasing the incidence of the disease.

The quarantine measures were also effective, with a reduction of 58.3% in cases, showing a significant impact in controlling the spread, particularly in high-risk areas.

The public health campaign had a 50% reduction in cases, which is effective but less impactful compared to vaccination and quarantine measures.

Overall, vaccination proved to be the most successful strategy in reducing disease cases, followed by quarantine and public health campaigns. These results suggest that a combination of these strategies could provide a comprehensive approach to controlling and mitigating the spread of infectious diseases.

### Question 10
a) Based on the trend, the number of cases in Week 5 is likely to decrease. Factors influencing this trend include the effectiveness of control measures like quarantine and face masks, and the reduced number of locations affected.

b) The most likely source of the initial outbreak is the Market. Evidence includes the fact that all initial cases reported visiting the Market, and environmental tests showed contaminated surfaces there.

## Unit 2: Topic 2: Infectious disease and epidemiology

c) The most probable mode of transmission is airborne. This is suggested by the symptoms (cough and fever) and the spread to various enclosed locations like the school, gym, and office, where airborne particles can easily spread the disease.

d) The closure of the Market and increased sanitation efforts in Week 3 significantly reduced the number of cases from 50 to 15 in Week 4. The Week 4 measures, including quarantine and face mask mandates, further reduced cases, but the Market closure had the most immediate and substantial impact.

e) An additional strategy could be enhancing ventilation in public places. Improved air circulation reduces the concentration of airborne pathogens, which is effective based on the inferred transmission mode through contaminated air or surfaces.

www.ingramcontent.com/pod-product-compliance
Lightning Source LLC
LaVergne TN
LVHW072001080526
838202LV00064B/6815